# Cambridge Elements ≡

Elements in Geochemical Tracers in Earth System Science
edited by
Timothy Lyons
*University of California*
Alexandra Turchyn
*University of Cambridge*
Chris Reinhard
*Georgia Institute of Technology*

# MOLYBDENUM AS A PALEOREDOX PROXY

## *Past, Present, and Future*

Stephan R. Hlohowskyj
*Central Michigan University*

Anthony Chappaz
*Central Michigan University*

Alexander J. Dickson
*Royal Holloway, University of London*

CAMBRIDGE
UNIVERSITY PRESS

# CAMBRIDGE
## UNIVERSITY PRESS

University Printing House, Cambridge CB2 8BS, United Kingdom

One Liberty Plaza, 20th Floor, New York, NY 10006, USA

477 Williamstown Road, Port Melbourne, VIC 3207, Australia

314–321, 3rd Floor, Plot 3, Splendor Forum, Jasola District Centre,
New Delhi – 110025, India

103 Penang Road, #05–06/07, Visioncrest Commercial, Singapore 238467

Cambridge University Press is part of the University of Cambridge.

It furthers the University's mission by disseminating knowledge in the pursuit of education, learning, and research at the highest international levels of excellence.

www.cambridge.org
Information on this title: www.cambridge.org/9781108995283
DOI: 10.1017/9781108993777

© Stephan R. Hlohowskyj, Anthony Chappaz, and Alexander J. Dickson 2021

First published 2021

*A catalogue record for this publication is available from the British Library.*

ISBN 978-1-108-99528-3 Paperback
ISSN 2515-7027 (online)
ISSN 2515-6454 (print)

# Molybdenum as a Paleoredox Proxy

## Past, Present, and Future

Elements in Geochemical Tracers in Earth System Science

DOI: 10.1017/9781108993777
First published online: August 2021

Stephan R. Hlohowskyj
*Central Michigan University*

Anthony Chappaz
*Central Michigan University*

Alexander J. Dickson
*Royal Holloway, University of London*

**Author for correspondence:** Stephan R. Hlohowskyj, hloho1sr@cmich.edu

**Abstract:** Molybdenum (Mo) is a widely used trace metal for investigating redox conditions. However, unanswered questions remain that concentration and bulk isotopic analysis cannot specially answer. Improvements can be made by combining new geochemical techniques to traditional methods of Mo analysis. In this Element, we propose a refinement of Mo geochemistry within aquatic systems, ancient rocks, and modern sediments through molecular geochemistry (systematically combining concentration, isotope ratio, elemental mapping, and speciation analyses). Specifically, to intermediate sulfide concentrations governing Mo behavior below the "switch-point" and dominant sequestration pathways in low-oxygen conditions. The aim of this work is to (1) aid and improve the breadth of Mo paleoproxy interpretations by considering Mo speciation and (2) address outstanding research gaps concerning Mo systematics (cycling, partitioning, sequestration, etc.). The Mo paleoproxy has potential to solve ever-complex research questions. By using molecular geochemical recommendations, improved Mo paleoproxy interpretations and reconstruction can be achieved.

**Keywords:** molybdenum, redox, paleoproxy, molecular geochemistry, isotopes

ISBNs: 9781108995283 (PB), 9781108993777 (OC)
ISSNs: 2515-7027 (online), 2515-6454 (print)

# Contents

# 1 Introduction

Molybdenum (Mo, $Z = 42$, $A_r = 94.95$ g mol$^{-1}$) has been recognized as a versatile trace metal for investigating paleoredox settings ever since its chemical behavior in natural systems was described (e.g., Goldschmidt, 1954). It has a high degree of chemical reactivity – both solid and dissolved phases – across a wide range of redox states leaving distinct geochemical signatures related to the depositional environment (Helz et al., 1996). As a paleoredox proxy, Mo has contributed to major breakthroughs regarding the ancient ocean and atmospheric chemistry of the early Earth (e.g., Anbar et al., 2007). Yet, despite considerable effort, the specifics on Mo cycling between dissolved, particulate, and solid phases are still controversial and actively debated (Chappaz et al., 2014; Wagner et al., 2017; Dahl et al., 2017; Vorlicek et al., 2018; Helz & Vorlicek, 2019). Molybdenum is widely distributed across the surface of the Earth, occurring in trace amounts within the crust, while juxtaposed as a highly concentrated transition metal in the ocean (Collier, 1985). It occupies a large range of oxidation states (–IV to +VIII) with +IV, +VI most commonly found on the Earth's surface. Additionally, Mo has seven naturally occurring stable isotopes (A ~ 92, 94, 95, 96, 97, 98, 100) with relatively similar abundances (i.e., ~10–25%). In the lithosphere, Mo can be concentrated up to weight percent in ore porphyry deposits, igneous bodies, magma contacts, or residual melts (Fig. 1). However, the disseminated average upper crustal (non-ore deposits) concentrations of Mo range from 1 to 3 ppm and are associated with neoformation of solid-phase minerals such as powellite (CaMo(VI)O$_4$), wulfenite (PbMo(VI)O$_4$), or weathering products of molybdenite (Mo(IV)S$_2$) (Ross & Sussman, 1955; Turekian & Wedepohl, 1961; Wedepohl, 1971; Bertine & Turekian, 1973; Erickson, 1973; Emerson and Huested, 1991). In the hydrosphere, Mo is supplied by oxidative weathering and hydrolysis of primary minerals (e.g., Mo(IV)S$_2$), to form the highly soluble – in pH 6 to 8 waters – oxyanion molybdate (Mo(VI)O$_4^{2-}$). In riverine and lacustrine systems, total molybdenum ($\Sigma[\text{Mo}]$) averages ~5–10 nM, with a range in some cases up to ~80 nM (Chappaz et al., 2008; Rahaman et al., 2010; Miller et al., 2011; Reimann & de Caritat, 2012; Glass el al., 2013). Across the oxygenated oceans Mo is generally ubiquitous (~105 nM) with an average residence time of ~0.44 to 0.8 Ma (Fig. 1) (Collier, 1985; Miller et al., 2011; Nakagawa et al., 2012). The distinct concentration contrast of solid-phase crustal $\Sigma[\text{Mo}]$ and long-lived aqueous-phase ($\Sigma\{\text{Mo}\}$) has fundamental redox and pH interpretations, especially when $\Sigma[\text{Mo}]$ is enriched ($\gg$ 1 ppm) in modern sediments or in the geological record. As a paleoredox proxy, Mo has proven to be an invaluable "forensic" tool for geochemists, especially to

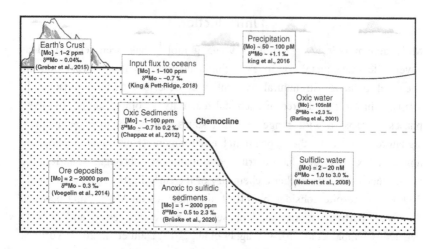

**Figure 1** Reservoirs of [Mo] on the surface of the Earth. Each box represents the approximate range of concentrations possible for dissolved or solid-phase Mo respectively (concentration average data from Bertine, 1972; Collier, 1985; Scott & Lyons, 2012).

identify strongly reducing episodes in sedimentary strata. Characterizing its chemical behavior in rocks, sediments, and pore waters has helped define the chemical composition of earth's early oceans (e.g., Lyons et al., 2009, Chappaz et al., 2017), identify changes in atmospheric oxygen concentration (e.g., Lyons et al., 2014), measure pervasiveness of oceanic redox conditions and total sulfide ($\Sigma S(-II)$) (e.g., Adelson et al., 2001), and understand early ocean productivity during the proliferation of multicellular life (e.g., Anbar & Knoll, 2002). Additionally, the study of natural Mo reservoirs has significantly advanced the fields of biogeochemistry and astrobiology, in part because Mo is essential for life. Nitrogen fixation and biological nitrate reduction rely on Mo as a cofactor in nitrogenase and nitrate reductase enzymes. Both enzymes are critical to the nitrogen cycle, and some research suggests they could be responsible for the first stages of life's diversity around 2.1 Ga (Glass et al., 2009; Boyd et al., 2011). Interestingly, Reinhard et al. (2013) suggest that Mo enrichment in rocks from sulfidic depositional environments across stable and long timescales (e.g., mid-Proterozoic) reflects Mo–N colimitation in the surface ocean. Their model implies a strong control on carbon and oxygen cycling via "bioinorganic feedbacks" related to redox-sensitive metals (i.e., Mo) and therefore highly influential but small seafloor regions could control biological cycling (Reinhard et al., 2013).

Until recently, the roles of organic matter (OM) associations and microbiological interactions as controls on Mo behavior have been difficult to quantify, compared to Mo–mineral associations (Wichard et al., 2009; Dahl et al., 2017; Wagner et al., 2017; Ardakani et al., 2018; King et al., 2018; Dickson et al., 2019). However, in highly productive systems, OM associations appear to have potential as a significant contributor to Mo enrichment. Molybdenum–mineral associations, however, have long been recognized as clear indicators of redox setting when investigating Mo in both isotopic signature and chemical composition (Helz et al., 1996; Arnold et al., 2004; Tribovillard et al., 2006; Chappaz et al., 2014; Scholz et al., 2017). For example, Scott et al. (2008) demonstrated the inorganic enrichment of $\Sigma[Mo]$ recorded ancient oceanic and atmospheric redox history in shales and indicated an increase in oxidative weathering driving Mo to the oceans, where a reducing redox gradient forced burial. This and other seminal works (e.g., Siebert et al., 2005; Anbar et al., 2007; Wille et al., 2007; Wille et al., 2008) helped improve our understanding concerning such important Earth events as the age of the first "whiffs" of oxygen in the atmosphere, the rise of photosynthetic life, Archean paleoredox conditions, and hydrogen sulfide–induced mass extinction events.

Isotopic signatures of Mo ($\delta^{98}Mo$) considerably improved the utility of the paleoproxy by adding the capability of fingerprinting mineral pathways to Mo deposition, and the ability to trace variations in the redox-sensitive burial fluxes of Mo at a global scale (Barling et al 2001; Siebert et.al 2003). Further applications have included insights into oceanic circulation, extinction events, anthropogenic sources, and, indirectly, large-scale glaciation (e.g., Pearce et al., 2008; Chappaz et al., 2012; Proemse et al., 2013; Chen et al., 2015; Kendall et al., 2015; Zhou et al., 2015; Dickson et al., 2017). Combined with concentrations, isotopic signatures have also resulted in the development of conceptual models of Mo cycling and speciation in water and sediments (e.g., Neubert et al., 2008; Helz et al., 2011; Nägler et al., 2011; Scholz et al., 2017). Yet, many models are derived from thermodynamic predictions with few empirical validations and/or limited observations of in situ Mo speciation (Erickson & Helz, 2000). This has led to a disparity between model predictions and empirical results (e.g., Dahl & Wirth, 2017). To overcome this discrepancy, Mo models require empirical measurements of in situ speciation and associated isotopic signature to validate parameters and/or proposed reactive pathways.

Herein, we present a synthesis of the new insights into the Mo paleoproxy offered by molecular geochemistry. We explore the environmental controls on Mo-redox coupling, the details of Mo cycling in aquatic systems, Mo enrichments in sediments, and new models for Mo cycling in natural systems. We highlight the potential strengths and limitations of molecular geochemistry, while providing an update to the latest research, methods, and techniques.

Finally, we conclude with suggestions on geochemical terminology, defined molecular geochemical processes, and future research directions.

## 2 Refining Redox Conditions

In the following sections, Mo reactions and processes are divided into distinct redox zones within an aquatic system (i.e., water column and sediments). Thus, before continuing, it is vital to provide clear definitions of terminology used to characterize these redox conditions (Fig. 2). Building upon the redox scheme proposed by Canfield and Thamdrup (2009), the term "oxic" hereafter is defined as a zone where aerobic respiration is dominant, and oxygen is the major electron acceptor (Fig. 2). The term "suboxic" is commonly used in geochemical literature, yet it has no definitive consensus, with varied meanings across the field (Canfield and Thamdrup, 2009). Therefore, the term "suboxic" will only be used as an analogy when unavoidable to indicate nitrogenous, manganous, and ferruginous zones (Fig. 2). The intermediate redox zones (i.e., nitrogenous, manganous, ferruginous) exist when nitrate ($NO_3^-/NO$), manganese ($Mn(IV)/Mn(II)$), and iron ($Fe(III)/Fe(II)$) redox couples respectively dominate, and neither $O_2$ nor $\Sigma S(-II)$ can be measured. Finally, the "sulfidic" zone is

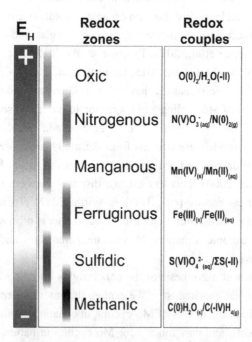

**Figure 2** Redox ladder with defined redox zones and corresponding redox couples

defined as the region where measurable $\Sigma S(-II)$ exists, after iron reduction has occurred and sulfate ($SO_4{}^{2-}$) reduction dominates. In the geochemical literature, "anoxic" and "euxinic" are often used to describe the presence of $\Sigma S(-II)$ in either the pore water or the water column, respectively. Unfortunately, the term "anoxia" is imprecise for the evaluation of molecular geochemistry, since it can span multiple redox zones and potentially a large range of Mo isotopic signatures. Therefore, the term "anoxic" is avoided, since it does not adequately describe the dominant redox chemistry at work. Similarly, "euxinic" is commonly used as an aquatic condition where abundant free $\Sigma S(-II)$ is present in the water column (e.g., Lyons et al., 2009) and hereafter is considered equivalent to the "sulfidic" zone.

# 3 Molybdenum Speciation

## 3.1 The Oxic Redox Zone

In the oxic environment, several Mo species can be simultaneously present: as a soluble anion (e.g., $Mo(VI)O_4{}^{2-}$, $Mo(VI)O_3$), as cations adsorbed to mineral surfaces (e.g., $Mo\text{-}FeOOH$, $Mo\text{-}MnOH_2$), as solid amorphous minerals (e.g., $Fe_xMo_yOH$, $Mn_xMo_yOH$), and associated with organic matter (e.g., $Mo\text{-}OM$) (Fig. 3; Helz et al., 1996; Wasylenki et al., 2008; Chappaz et al., 2014; Dahl et al., 2017). Each of these species in the natural environment plays an important role in the cycling and (bio)availability of Mo to both solid and dissolved reservoirs. Dissolved Mo ($\Sigma\{Mo\}_{(aq)}$) in oxic waters is often considered less reactive and conceptually only becomes a particle reactive element under strongly reducing conditions (e.g., Helz et al., 1996). However, $Mo(VI)O_4{}^{2-}{}_{(aq)}$ has been demonstrated to contribute to Mo shuttling, sequestration, and overall availability of $\Sigma\{Mo\}_{(aq)}$ in some systems (Siebert et al., 2003; Glass et al., 2013; Smedley & Kinniburgh, 2017). For example, oxic zone adsorption to manganese (Mn) and iron (Fe) oxyhydroxides, and associations to OM can actively shuttle Mo to the sediment–water interface (SWI) where enrichment occurs, often in the presence of sulfidic pore water (Scott & Lyons, 2012).

Molybdenum's affinity to Mn occurs most commonly in fully oxic ocean basins, where nodules can form. Manganese oxides occur in the water column via precipitation, hydrolysis, and oxidation of $Mn(II)_{(aq)}$ to $Mn(IV)_{(s)}$ (Barling & Anbar, 2004; Kendall et al., 2017), leading to Mo adsorption on $Mn(IV)O_2$ surfaces (Wasylenki et al., 2011). This process facilitates particulate $Mn\text{-}Mo$ shuttling to the SWI, causing a positive correlation of $Mn(IV)O_2$ to $\Sigma[Mo]_{(s)}$ in suspended particles, while negatively correlating dissolved Mn and $\Sigma\{Mo\}_{(aq)}$ (Berrang & Grill, 1974). Manganese shuttling is most prevalent in zones of oceanic upwelling (e.g., west coast of South America), where lower-oxygen-content waters reach the

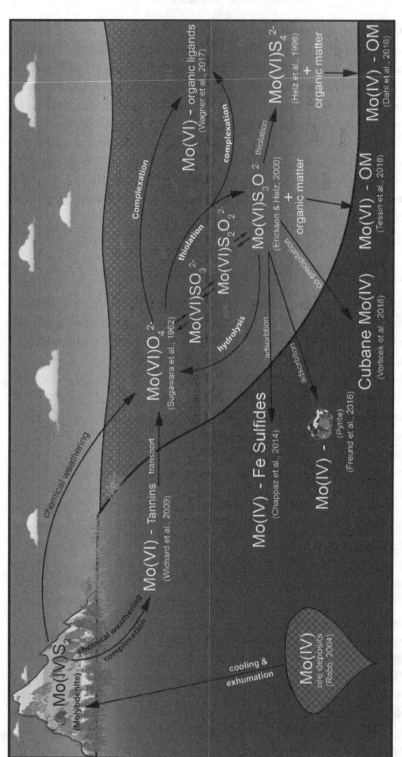

**Figure 3** Generalized redox pathways for Mo in aquatic systems, including solid-phase sequestration.

surface, oxidizing Mn(II) species, facilitating Mn(IV)$O_2$ precipitation, and promoting Mo adsorption (Bertine & Turekian, 1973; Calvert & Price, 1983; Seralathan and Hartmann, 1986). Molybdenum can adsorb in significant amounts to Fe oxides and hydroxides such as goethite (FeO(OH)) at lower pH (e.g., $\leq 6$). However, this effect is more common in surficial sediments, soils, and pore waters than in oceanic systems (Goldberg et al., 1996). Clay minerals (e.g., illite, smectite) can also adsorb Mo most effectively at pH 5–6 but kinetically slows to a stop at pH $\geq 8$ (Goldberg et al., 1996; Goldberg & Forster, 1998).

Molybdenum mineral adsorption shuttling is well documented in low-oxygen-concentration ($[O_2]_{(aq)} < 5$ μM) systems (e.g., Santa Monica Basin), where significant amounts of Mo, Fe, and Mn coexist (Shaw et al., 1990). In seasonally oxic ocean basins (e.g., Saanich Inlet), where $[O_2]$(aq) can reach 0 μM during occasional manganous-to-ferruginous reducing events, mineral adsorption and shuttling has been found to be a primary Mo enrichment mechanism (Crusius et al., 1996). These pathways, however, do not appear to generate a reduction of Mo (i.e., Mo(VI) $\rightarrow$ Mo(IV)) within the water column but may catalyze it in reducing pore waters and/or sediments. Isotopically, Mo mineral oxide adsorption does cause a well-documented isotopic fractionation to lower $\delta^{98}$Mo values, observed in both laboratory experiments and natural samples (Barling & Anbar, 2004; Reitz et al., 2007; Wasylenki et al., 2008, 2011; Goldberg et al., 2009; Kashiwabara, et al., 2011). Molybdenum and OM concentrations have been commonly observed to covary linearly together in the geologic record and are a potentially important primary sink for sedimentary rocks (i.e., shales) and ocean sediments (Brumsack & Gieskes, 1983; Holland, 1984; Brumsack, 1986; Algeo & Lyons, 2006). The mechanisms explaining the variation of slopes related to the positive Mo-OM correlation are still poorly understood and actively debated, with a variety of proposed interactions between Mo and OM or basin configurations during deposition. For example, Helz et al. (2019) calculated the hypothetical influence of OM on Mo, concluding that correlations found in the geologic record are due to simultaneous OM and Mo sequestration pathway(s). These pathways act independently from each other, concentrating both Mo and OM, which by happenstance linearly correlate. However, laboratory-based results refute these predictions, showing that Mo readily associates with organic acids (e.g., Wagner et al., 2017) and particulate OM (e.g., Dahl et al., 2017; Dickson et al., 2019). In natural samples, Mo has been observed to associate with dissolved organic carbon (DOC) and/or low-molecular-weight OM (Brumsak & Gieskes, 1983; Scholz et al., 2018; Tessin et al., 2019; Greaney et al., 2020). For instance, in organic-rich sections of diatomaceous sediments from Namibia, Mo is associated with fulvic and humic acids (Calvert & Morris, 1977). Various explanations have been put forth to explain these Mo-OM data, ranging from Mo sorption on particulate organic matter (POM), complexation of Mo to unknown organic molecules, bonding to humic acids (e.g., Szilagyi, 1967; Nissenbaum & Swaine, 1976), or high biological productivity causing Mo(VI) reduction to Mo(V) via an OM electron donor (e.g., Brumsak & Gieskes, 1983). However, there is still no consensus on the dominant mechanism(s) relating Mo to OM in sediments (Algeo & Lyons, 2006). Indeed, the sheer complexity of OM and the observed associations with Mo point to a likely multi-pathway shuttle dependent on OM-type and dominant redox conditions. Isotopically, few direct measurements of natural $\delta^{98}$Mo from organic associations are published to date; however, results from Dickson et al. (2019) suggest a lack of fractionation from the bulk $\delta^{98}$Mo signal. Yet, hypothetically a Mo-OM association should lead to lower isotopic fractionation when Mo transitions to solid phase(s), similar to biological mediated fractionation (e.g., Wasylenki et al. 2007; Zerkle et al., 2011). This complicates the already intricate isotope system, with possible unknown Mo-organic contributions in geologic samples (Kendall et al., 2017).

In general, burial and enrichment of Mo in oxic redox regimes is a function of both mineral and organic pathways shuttling dissolved and solid- (particulate-) phase Mo to the SWI. However, most literature has focused on mineral phases (e.g., Mo-FeO(OH), Mo-MnO$_2$) and not OM, leading to a significant research gap. This is partly due to practical and logistic impediments to direct measurements of Mo below the SWI while preserving speciation and depositional redox state. Merely sampling pore waters can alter Mo speciation and oxidation state, even under the best field circumstances, by introducing atmospheric O$_2$. Additionally, in an ideal oxic system, the top centimeters of the SWI are likely dominated by diffusive fluxes of $\Sigma\{Mo\}_{(aq)}$, while Mo solid-phase transitions occur deeper in the sediment (Crusius et al., 1996). Attempting to capture this redox profile and original depositional conditions is problematic, since Mo below the first few centimeters of the SWI can have multiple transport or burial processes occurring at the same time.

## 3.2 The Nitrogenous, Manganous, and Ferruginous Redox Zone(s)

Studies of these redox zones are limited in the literature, since they are not well documented in natural samples, are difficult to analyze in situ, and lack agreed-upon methods to capture Mo speciation and oxidation state without influencing results. There is strong environmental evidence that very low, or zero bottom-water oxygen ($[O_2]_{(aq)} < 1$ μM) concentrations promote formation of measurable authigenic solid-phase Mo minerals (Crusius et al., 1996). However, little data exists on water-column redox zone transitions (from oxic to ferruginous) and the subsequent effect on Mo speciation (Scholz et al., 2017). In continental margin sediments with oxic water columns and small amounts of sulfide in pore waters – well below the SWI – Mo solid-phase enrichments can range up to one order of magnitude compared to crustal levels (Hardisty et al., 2018). These Mo enrichments demonstrate a reliable Mo shuttle to the SWI, with sequestration enhanced by nitrogenous, manganous, and ferruginous redox gradients until some sulfide is available (Scholz, et al., 2017). Furthermore, Scott and Lyons (2012) imply that although these intermediate redox regimes are less common, they still contribute significantly to solid-phase Mo concentrations. However, in an SWI profile under an oxic water column, the transition across these redox regimes would account for only very thin layers (10–20 mm) of pore space where redox reactions could occur (Pederson et al., 1989; Zheng et al., 2000). In these layers, increased solubility of iron within the ferruginous zone promotes a downward flux due to Mo-mineral precipitation deeper in the sediment profile (Hardisty et al., 2018). These redox gradients might provide kinetically favorable conditions for mineral Mo coprecipitation of poorly crystalline to amorphous materials (Vorlicek et al., 2018). Furthermore, Mo-OM associations have been found in sediments deposited within these intermediate redox zones (Crusius et al., 1996). In some specialized systems, organically bound Mo may be more effectively buried than mineral pathways, either through shuttling to an SWI with sulfidic pore water or becoming an OM bound sink before any sulfide mineral precipitation can occur (Dahl et al., 2017; Scott et al., 2017). In sediments the overall solubility of $\Sigma\{Mo\}_{(aq)}$ decreases rapidly in pore waters as oxygen is consumed and continues to decrease with depth as $\Sigma[Mo]_{(s)}$ gradually increases, presumably from precipitation of authigenic solid phases. However, it is unclear if and when solid-phase Mo(VI) undergoes reduction to Mo(IV) during burial or diagenesis in these systems. In several cases, Mo oxidation state has been reported as mixed between Mo(VI) and Mo(IV), with potentially intermediate redox zones during deposition (Dahl et al., 2017). It is likely that, in intermediate redox zones, Mo tends to remain in an oxidized state, Mo(VI), and only undergoes a coordination change (e.g., tetrahedral to trigonal), promoting a solid-phase precipitation as Mo solubility decreases. Once in a solid phase, Mo is more susceptible to reduction to Mo(IV), especially in the presence of sulfide commonly found in pore waters below the SWI.

The difficulty of studying such redox zones stems from their sensitivity to oxygen or sulfide. To properly examine the molecular geochemistry of these redox zones, collected samples need to be flash frozen at temperatures $\leq -80\ °C$ and maintained in an anoxic nonreactive (e.g., nitrogen) atmosphere to completely halt redox reactions. These methods need to be maintained consistently during sample collection, preparation, and even analysis to avoid any introduced bias. Additionally, in the natural environment there are few regions that are consistently stable within these redox zones, with most localities being seasonal or existing as end-members of oxic to sulfidic (Berrang & Grill, 1974). Thus, the best picture of molecular geochemistry in nitrogenous, manganous, and ferruginous redox zones is inferred from sediment core profiles when both pore-water Mo and solid-phase Mo can be collected, preserved, and analyzed.

## 3.3 The Sulfidic Zone(s)

In sulfidic redox zones, Mo enrichments follow a predictable pattern, depending on $\Sigma S(-II)$ free in the water column and/or pore waters below the SWI (Scott and Lyons, 2012). This is generally related to an increased affinity for solid phases as solubility decreases because of Mo thiolation. Contrary to other Mo shuttles, natural sulfidic waters (pH 6–8) thermodynamically favor thiolation of dissolved $Mo(VI)O_4^{2-}$ and promote solid-phase sequestration in sediments (Helz et al., 1996; Erickson & Helz, 2000; Chappaz et al., 2014). Thiomolybdates are known to easily concentrate in solid phases at orders of magnitude higher than average crustal backgrounds (e.g., 10–100 ppm) via reactive pathways involving OM and sulfide mineral phases (Chappaz et al., 2014; Dahl & Wirth, 2017; Wagner et al., 2017). Since many interpretations of Mo redox sensitivity exist within the sulfidic redox regime (e.g., Helz et al., 1996; Dahl et al., 2010; Chappaz et al., 2014), hereafter it is defined as Mo thiolation when $\Sigma S(-II) \geq 100$ μM, often coinciding with formation of sulfide mineral phases. There is evidence that points to an important and telling pathway dictated by Mo-S-Fe coprecipitation depending upon $\Sigma S(-II)$ and pH (Helz et al., 2011; Vorlicek et al., 2018; Helz & Vorlicek, 2019). Yet, Mo sulfide sensitivity is also detailed in numerous studies quantifying other specifics of Mo cycling, partitioning, burial, reaction kinetics, and isotopic signature, during times where water column $\Sigma S(-II) \geq 100$ μM (Helz et al., 1996; Tribovillard et al., 2006; Scott & Lyons, 2012; Chappaz et al., 2014). This large body of research exists likely due to the ease of measuring significant (e.g., 2–10 orders of magnitude greater than crustal values) Mo enrichments in sediments and rocks, which consistently occur when $\Sigma S(-II) \geq 100$ μM.

A common misconception of $Mo(VI)O_4^{2-}$ thiolation is that it is accompanied by an immediate reduction of Mo(VI) to Mo(IV). Thiolation, however, describes only substitution, and not a reduction reaction. Molybdenum reduction is likely promoted as solubility decreases, causing solid-phase formation of Mo species (Erickson & Helz, 2000). However, several studies show that Mo oxidation state and associated reductive pathways are not simplistic as previously thought (Dahl et al., 2013). Research measuring oxidation state using

X-ray Absorption Fine Structure (XAFS) has demonstrated that solid-phase Mo can exist in multiple oxidation states associated with different mineral phases, OM, or amorphous solids (Ardakani et al., 2016; Tessin et al., 2019; Vorlicek et al., 2018). In systems with mixed Mo oxidation states, the ratio of Mo(VI) to Mo(IV) might represent a gradient related to $\Sigma S(-II)$ availability, total iron ($\Sigma Fe$) availability, Mo species, and total organic carbon (TOC) content. In rare cases, Mo(V) has been reported; however, there is little evidence to suggest that Mo(V) is a stable, long-lived, and thermodynamically favorable state to consistently be preserved in sediments (Wang et al., 2011). In seasonally sulfidic depositional basins such as the Saanich Inlet or the Namibian Shelf, solid-phase Mo enrichments and water column thiolation are well documented; however, the dominant sequestration mechanism likely varies depending on the redox seasonality (Brongersma-Sanders et al., 1980). This also occurs in so-called "weakly euxinic" zones (20 $\mu$M $\leq \Sigma S(-II) \leq$ 100 $\mu$M), where intermediate thiolated species can dominate and Mo is enriched in solid phases (Berrang & Grill, 1974).

## 4 Molybdenum Isotopes

Molybdenum isotopes can characterize the end-member compositions of Mo adsorbed or bound to oxic minerals (e.g., birnessite or haematite), with values ~2.2–3‰ lower than seawater (Barling and Anbar, 2004; Wasylenki et al., 2007; Goldberg et al., 2009). In sulfidic redox regimes when $\Sigma S(-II) \geq$ 100 $\mu$M, heavy Mo isotopic signatures in sediments are interpreted to represent the thiolation of dissolved Mo toward an estimated equilibrium fractionation for tetrathiomolybdate ($MoS_4^{2-}$) approaching ~0.5 ‰ lower than coeval basin seawater (Nägler et al., 2011). In practice, the rapid sequestration of dissolved Mo into sediments after thiolation leads to near-quantitative removal in some settings and a sedimentary isotope composition approaching that of open-ocean seawater (today ~2.3‰). Open-ocean sulfidic sediments have compositions approaching seawater but are usually offset by ~0.5–0.9 ‰ due to nonquantitative removal (Poulson Brucker et al., 2009; Hutchings et al., 2020; Sweere et al., 2021) or mixing with isotopically lighter ferruginous or manganiferous sedimentary phases (e.g., Siebert et al., 2006; Scholz et al., 2017).

However, the interpretation of Mo isotope signatures is complicated in intermediate redox zones (nitrogenous, manganous, ferruginous, weakly sulfidic) when oxygen ($[O_2]_{(aq)} \leq 5 \mu$M) and free sulfide ($\Sigma S(-II) \leq 100 \mu$M) are low. In these settings, equilibrium fractionation signatures of intermediate thiomolybdates ($MoO_xS_{4-x}^{2-}$) are several per mil lower than seawater when preserved (Tossell, 2005; Azrieli-Tal et al., 2014). Sedimentary Mo isotopes in ferruginous or nitrogenous conditions are further complicated by mixing of sulfide, ferruginous, and manganiferous phases, which leads to highly variable isotopic signatures (e.g., −1‰ to 2.3‰) (Poulson Brucker et al., 2009; Hutchings et al., 2020; Sweere et al., 2021). However, the lack of speciation data for Mo buried in intermediate redox-state conditions makes the interpretation of isotopic signatures difficult. Depending on the dominant thiolated species, the isotopic signature of dissolved Mo could be vastly different from typical seawater values, especially in systems with $\Sigma S(-II) \leq 100$ μM, as specific species are preferentially sequestered to solid phases over others. Additionally, the already difficult $\delta^{98}$Mo system has an unknown isotopic contribution from Mo-OM associations (Kendall et al., 2017) and might be affected by mixing with allochthonous organic-rich fluids following geological burial (e.g., Ardakani et al., 2020). Within the oxic setting, the well-known Mn–Fe-oxyhydroxide shuttle can accelerate enrichment of Mo into seafloor sediments (Algeo & Tribovillard, 2009). However, as Mo is supplied to the sediments, if the redox setting at or below the sediment–water interface is sufficiently sulfidic (e.g., beyond the switch-point [Helz et al., 1996]), Mo is removed from the aqueous phase into the solid phase and generally not recycled in the water column, altering its isotopic profile. Transition between dissolved and solid phases of Mo indeed contributes to isotope fractionation, as observed in the shallow Black Sea (Nägler et al., 2005, 2011). Additionally, understanding these modern systems (e.g., Cariaco Basin) can provide potential exemplars of ancient ocean basins and thus help to deconvolute the Mo geologic isotope record as a whole (e.g., Brüske et al., 2020; Noordman et al., 2015).

## 5 Research Gaps

### 5.1 Limitations of Current Approaches

The current arsenal of methods available to geochemists is vast. However, most rely on the combination of two methods: inductively coupled plasma mass spectrometry (ICP-MS) for concentration, and multi-collector inductively coupled plasma mass spectrometry (MC-ICP-MS) for isotope measurements. Rapid simultaneous measurements of multiple trace metal concentrations (~μg/L) become feasible only with widespread availability, reduced cost, and user-friendliness of ICP-MS. This technology allows measurements of additional redox-sensitive trace metals such as iron (Fe), manganese (Mn), vanadium (V), chromium (Cr), and uranium (U) in rocks and sediments at concentrations orders of magnitude (e.g., $\leq$ 1 μg/kg) lower than those obtained by conventional sequential extraction methods (Tribovillard et al., 2006). So-called nontraditional (i.e., Fe, Mn, Mo, V, Cr, U) isotopic measurements via MC-ICP-MS better constrain redox chemistry of aquatic systems. When combined, these tools impart a considerable amount of information for the user; however, they are not without limitations.

Historically, geochemical measurements of redox-sensitive trace metals were confined to bulk geochemistry of rocks, water, or sediments, often using sequential extraction, ultraviolet visible light absorption, or flame atomic absorption analysis (Goldschmidt 1954; Kuroda & Sandell, 1954; Wedepohl 1971). Although these techniques have helped interpret the redox setting within ancient oceans (e.g., Anbar & Knoll, 2002), they often do not uniquely quantify the specific mechanisms enriching trace metals. With ICP-MS and MC-ICP-MS, geochemical interpretations have improved drastically, allowing major discoveries on the paleoredox history of Earth to be made (e.g., Anbar et al., 2007). However, the shortfall of these bulk geochemical methods resides in the inability to resolve specific mechanistic trace metal pathways, due to sample size requirements and the need to homogenize materials before analysis. Bulk measurement requirements cause ambiguity in trace metal sequestration and loss of information when measuring geologic materials. Despite this, these methods are rapid and offer a low-cost "first pass" to identify key intervals in a geologic record or sediment archive.

## 5.2 Molecular Geochemistry

Molecular geochemistry (i.e., combining analyses of concentration, isotope ratio, elemental mapping, and speciation) of trace metals has gained momentum among geochemists by providing more-nuanced geologic interpretations identifying specific species, reactions, and sequestration pathways (e.g., Ardakani et al., 2020). Specifically, it can be used to directly measure oxidation state, dominant species, bonding distance, coordination, and adsorptive affinity of trace metals. However, until recently these methods have been underused by geochemists due to lack of general awareness, little access to advanced training, and misunderstanding of costs. The lack of use of molecular geochemistry is unfortunate, since foundational works such as Siebert et al. (2003) demonstrated the importance of recognizing that fractionation of isotopic values could be tied to coordination change of tetrahedral to octahedral, outweighing a simple model of kinetic fractionation of an initial isotope pool. Still, these practices are continuing to spur a geochemical renaissance by improving and refining depositional settings, defining precise water column redox conditions, and identifying paleo pH.

Molecular geochemistry is a rapidly evolving subfield; however, the primary obstacle to its continued advancement is the need for rigorous laboratory-based experimental data combined with innovation of existing methods coupled with "cutting-edge" analytical instrumentation (e.g., Vorlicek at al., 2018 and references therein). For example, high-energy X-rays (i.e., X-ray Absorption Fine Structure [XAFS]) can measure oxidation state and coordination number; nuclear scattering (i.e., neutron diffraction [ND]) can measure crystal structure and light isotopes; and nanometer-scale ionization (e.g., secondary ion mass spectrometry [SIMS]) can measure isotopic fractionation associated within single mineral matrices or OM complexes (Liang et al., 2008; Reich et al., 2013). However, before using these technologies, the geologic materials in question need to have a complete and comprehensive geochemical evaluation to gain the most value from the output molecular geochemical data. Thus, the best practice for studying Mo systematics uses a multipronged approach of bulk concentrations and isotopic signatures coupled with molecular geochemistry to paint a complete paleoredox picture.

## 6 Future Research Directions

Molybdenum is a powerful geochemical tool that has evolved significantly in use during the past fifty years. Although many important scientific questions were answered using Mo systematics (i.e., total concentrations and isotopic signatures), these alone are not enough to answer many current geochemical research questions. Most notably, the Mo paleoproxy falls short in intermediate redox conditions (nitrogenous, manganous, ferruginous), where multiple dissolved and solid-phase sequestration pathways appear to be acting together on the bulk geochemical signal of Mo. These redox zones remain a problem for the geochemical community due to the potential for misinterpretations or oversimplification in the geologic record. To improve the geochemist's ability to understand the dynamics of Mo cycling across all redox systems, training in and use of molecular geochemistry to measure species, oxidation state, coordination, bonding, and phase association are needed. By combining these state-of-the-art analytical techniques with traditional approaches, the possibility of fully understanding Mo geochemistry is within the reach of scientists. Additionally, molecular geochemistry is not limited to just the Mo paleoredox proxy; the usefulness of these methods applies to any redox-sensitive trace metal and, through continued research endeavors, will expand the ability of anyone to precisely describe depositional redox settings. The future geochemical frontier, therefore, lies in the ability to analyze trace metal molecular geochemistry in situ and across a wide variety of environmental conditions to build a robust geochemical knowledge base of different proxies.

# Recommended Reading

1. Smedley, P. L., & Kinniburgh, D. G. (2017). Molybdenum in natural waters: A review of occurrence, distributions and controls. *Applied Geochemistry*, 84, 387–432.

   Smedley and Kinniburgh (2017) provide the most extensive review of molybdenum in the natural world from solid to dissolved phases as it cycles in the environment.

2. Kendall, B., Dahl, T. W., & Anbar, A. D. (2017). The stable isotope geochemistry of molybdenum. *Reviews in Mineralogy and Geochemistry*, 82(1), 683–732.

   Kendall et al. (2017) summarize the major concepts in the use of molybdenum isotopes as a paleoproxy and provide an excellent summary of the current research gaps concerning isotope interpretation.

3. Dickson, A. J. (2017). A molybdenum-isotope perspective on Phanerozoic deoxygenation events. *Nature Geoscience*, 10(10), 721–726.

   Dickson (2017) covers the important and recent advances in isotope interpretations from the Phanerozoic and provides a useful synopsis complementing Kendall et al. (2017).

4. Helz, G. R., Miller, C. V., Charnock, J. M. et al. (1996). Mechanism of molybdenum removal from the sea and its concentration in black shales: EXAFS evidence. *Geochimica et Cosmochimica Acta*, 60(19), 3631–3642.

   Helz et al. (1996) is likely the most seminal and important paper concerning the use of molybdenum as a paleoproxy. In this article Helz et al. (1996) set the stage for studying the speciation of molybdenum in ancient systems.

5. Scott, C., Lyons, T. W., Bekker, A. et al. (2008). Tracing the stepwise oxygenation of the Proterozoic ocean. *Nature*, 452(7186), 456–459.

   Scott et al. (2008) is a seminal paper illustrating the use of the molybdenum paleoproxy to understand Earth's oxygenation and documents the power of molybdenum as a redox-sensitive tracer of past conditions.

6. Erickson, B. E., & Helz, G. R. (2000). Molybdenum (VI) speciation in sulfidic waters: stability and lability of thiomolybdates. *Geochimica et Cosmochimica Acta*, 64(7), 1149–1158.

   Erickson and Helz (2000) outline the chemistry of molybdenum speciation under sulfidic redox conditions and describe the kinetics of molybdenum thiolation to help interpret both modern and ancient sedimentary systems.

7. Scott, C., & Lyons, T. W. (2012). Contrasting molybdenum cycling and isotopic properties in euxinic versus non-euxinic sediments and sedimentary rocks: Refining the paleoproxies. *Chemical Geology*, 324, 19–27.

   Scott and Lyons (2012) identify the behavior of molybdenum across a range of redox conditions in both sediments and rocks. This paper illustrates the mechanics of molybdenum sequestration under reducing conditions and how geochemical signals can be interpreted for paleoredox reconstruction.

8. Algeo, T. J., & Lyons, T. W. (2006). Mo–total organic carbon covariation in modern anoxic marine environments: Implications for analysis of paleoredox and paleohydrographic conditions. *Paleoceanography*, 21(1), PA1016, DOI:10.1029/2004PA001112.

   Algeo and Lyons (2006) address the long-standing question of the role of organic matter in association with molybdenum commonly found in ancient rocks. This paper sets the stage for future research into organic matter dynamics relating to molybdenum.

9. Chappaz, A., Lyons, T. W., Gregory, D. D. et al. (2014). Does pyrite act as an important host for molybdenum in modern and ancient euxinic sediments? *Geochimica et Cosmochimica Acta*, 126, 112–122.

   Chappaz et al. (2014) outline the burial pathways for molybdenum sequestration in sediments, commonly thought to be dominated by pyrite precipitation. They highlight current hypotheses for possible molybdenum pathways and postulate on the potential of additional sequestration pathways.

10. King, E. K., & Pett-Ridge, J. C. (2018). Reassessing the dissolved molybdenum isotopic composition of ocean inputs: The effect of chemical weathering and groundwater. *Geology*, 46(11), 955–958.

    King and Pett-Ridge (2018) describe how Mo sourced from groundwater varies in concentration and isotopic composition compared to riverine inputs. This paper helps improve the understanding of isotopic cycling of Mo, strengthening the usefulness of the Mo paleoproxy by further constraining sources of variation.

11. Poulson Brucker, R. L., McManus, J., Severmann, S., & Berelson, W. M. (2009). Molybdenum behavior during early diagenesis: Insights from Mo isotopes. *Geochemistry, Geophysics, Geosystems*, 10(6), Q06010, DOI: https://doi.org/10.1029/2008GC002180.

    Poulson Brucker et al. (2009) show empirical measurements of Mo isotopic values across a wide range of redox regimes. This work is important in that it documents major marine sediment reservoirs of Mo while providing corresponding isotopic values.

12. Noordmann, J., Weyer, S., Montoya-Pino, C. et al. (2015). Uranium and molybdenum isotope systematics in modern euxinic basins: Case studies

from the central Baltic Sea and the Kyllaren fjord (Norway). *Chemical Geology*, 396, 182–195.

Noordmann et al. (2015) describe the importance of uranium geochemistry and its influence on Mo within sulfidic redox systems. Further, their research lays out a foundation concept of the importance of permanent stratification of a basin for the interpretation/reconstruction of paleoredox conditions.

13. Neely, R. A., Gislason, S. R., Ólafsson, M. et al.(2018). Molybdenum isotope behaviour in groundwaters and terrestrial hydrothermal systems, Iceland. *Earth and Planetary Science Letters*, 486, 108–118.

    Neely et al. (2018) outline the importance of lesser-studied Mo inputs to the oceanic pool: hydrothermal fluids. Their work demonstrates the need to include increasingly precise, although complex, considerations of all Mo inputs to the ocean budget, since the potential input of hydrothermal fluids is nontrivial.

# References

Adelson, J. M., Helz, G. R., & Miller, C. V. (2001). Reconstructing the rise of recent coastal anoxia; molybdenum in Chesapeake Bay sediments. *Geochimica et Cosmochimica Acta*, 65(2), 237–252.

Algeo, T. J., & Lyons, T. W. (2006). Mo–total organic carbon covariation in modern anoxic marine environments: Implications for analysis of paleoredox and paleohydrographic conditions. *Paleoceanography*, 21(1). DOI: https://doi.org/10.1029/2004PA001112.

Algeo, T. J., & Tribovillard, N. (2009). Environmental analysis of paleoceano-graphic systems based on molybdenum–uranium covariation. *Chemical Geology*, 268(3–4), 211–225.

Anbar, A. D., & Knoll, A. H. (2002). *Proterozoic ocean chemistry and evolution: A bioinorganic bridge? Science*, 297(5584), 1137–1142.

Anbar, Ariel D., Yun Duan, Timothy W. Lyons, et al. A whiff of oxygen before the Great Oxidation Event? *Science* 317, no. 5846 (2007): 1903–1906.

Ardakani, O. H., Chappaz, A., Sanei, H., & Mayer, B. (2016). Effect of thermal maturity on remobilization of molybdenum in black shales. *Earth and Planetary Science Letters*, 449, 311–320.

Ardakani, O. H., Hlohowskyj, S. R., Chappaz, A., et al. (2020). Molybdenum speciation tracking hydrocarbon migration in fine-grained sedimentary rocks. *Geochimica et Cosmochimica Acta*, 283, 136–148.

Ardakani, O. H., Sanei, H., Ghanizadeh, A., et al. (2018). Do all fractions of organic matter contribute equally in shale porosity? A case study from Upper Ordovician Utica Shale, southern Quebec, Canada. *Marine and Petroleum Geology*, 92, 794–808.

Arnold, G. L., Anbar, A. D., Barling, J., & Lyons, T. W. (2004). Molybdenum isotope evidence for widespread anoxia in mid-Proterozoic oceans. *Science*, 304(5667), 87–90.

Azrieli-Tal, I., Matthews, A., Bar-Matthews, M., Almogi-Labin, A., Vance, D., Archer, C., & Teutsch, N. (2014). Evidence from molybdenum and iron isotopes and molybdenum–uranium covariation for sulphidic bottom waters during Eastern Mediterranean sapropel S1 formation. *Earth and Planetary Science Letters*, 393, 231–242.

Barling, J., & Anbar, A. D. (2004). Molybdenum isotope fractionation during adsorption by manganese oxides. *Earth and Planetary Science Letters*, 217 (3–4), 315–329.

Barling, J., Arnold, G. L., & Anbar, A. D. (2001). Natural mass-dependent variations in the isotopic composition of molybdenum. *Earth and Planetary Science Letters*, 193(3–4), 447–457.

Berrang, P. G., & Grill, E. V. (1974). The effect of manganese oxide scavenging on molybdenum in Saanich Inlet, British Columbia. *Marine Chemistry*, 2(2), 125–148.

Bertine, K. (1972). The deposition of molybdenum in anoxic waters. *Marine Chemistry*, 1(1), 43–53.

Bertine, K. K., & Turekian, K. K. (1973). Molybdenum in marine deposits. *Geochimica et Cosmochimica Acta*, 37(6), 1415–1434.

Boyd, E. S., Anbar, A. D., Miller, S., et al. (2011). A late methanogen origin for molybdenum-dependent nitrogenase. *Geobiology*, 9(3), 221–232.

Brongersma-Sanders, M., Stephan, K. M., Kwee, T. G., & De Bruin, M. (1980). Distribution of minor elements in cores from the Southwest Africa shelf with notes on plankton and fish mortality. *Marine Geology*, 37(1–2), 91–132.

Brumsack, H. J. (1986). The inorganic geochemistry of Cretaceous black shales (DSDP Leg 41) in comparison to modern upwelling sediments from the Gulf of California. *Geological Society, London, Special Publications*, 21(1), 447–462.

Brumsack, H. J., & Gieskes, J. M. (1983). Interstitial water trace-metal chemistry of laminated sediments from the Gulf of California, Mexico. *Marine Chemistry*, 14(1), 89–106.

Brüske, A., Weyer, S., Zhao, M. Y., Planavsky, N. J., Wegwerth, A., Neubert, N., . . . & Lyons, T. W. (2020). Correlated molybdenum and uranium isotope signatures in modern anoxic sediments: Implications for their use as paleoredox proxy. *Geochimica et Cosmochimica Acta*, 270, 449–474.

Calvert, S. E., & Morris, R. J. (1977). Geochemical studies of organic-rich sediments from the Namibian Shelf. II. *Metal-organic associations. In M. Angel, ed., A Voyage of Discovery* (pp. 667–680). Pergamon, London.

Calvert, S. E., & Price, N. B. (1983). Geochemistry of Namibian shelf sediments. In E. Suess & J. Thiede, eds., *Coastal Upwelling Its Sediment Record. Part A: Responses of the Sedimentary Regime to Present Coastal Upwelling* (pp. 337–375). Springer, Boston.

Canfield, D. E., & Thamdrup, B. (2009). Towards a consistent classification scheme for geochemical environments, or, why we wish the term "suboxic" would go away. *Geobiology*, 7(4), 385–392.

Chappaz A., Glass, J. B., & Lyons, T. W. (2017). Molybdenum. In W. White (ed.), *Encyclopedia of Geochemistry*. Encyclopedia of Earth Sciences Series. DOI: https://doi.org/10.1007/978-3-319-39193-9_256-1.

Chappaz, A., Gobeil, C., & Tessier, A. (2008). Geochemical and anthropogenic enrichments of Mo in sediments from perennially oxic and seasonally anoxic lakes in Eastern Canada. *Geochimica et Cosmochimica Acta*, 72(1), 170–184.

Chappaz, A., Lyons, T. W., Gordon, G. W., & Anbar, A. D. (2012). Isotopic fingerprints of anthropogenic molybdenum in lake sediments. *Environmental Science & Technology*, 46(20), 10934–10940.

Chappaz, A., Lyons, T. W., Gregory, D. D., et al. Does pyrite act as an important host for molybdenum in modern and ancient euxinic sediments? *Geochimica et Cosmochimica Acta*, 126(2014), 112–122.

Chen, X., Ling, H. F., Vance, D., et al. (2015). Rise to modern levels of ocean oxygenation coincided with the Cambrian radiation of animals. *Nature Communications*, 6(1), 1–7.

Collier, R. W. (1985). Molybdenum in the Northeast Pacific Ocean 1. *Limnology and Oceanography*, 30(6), 1351–1354.

Crusius, J., Calvert, S., Pedersen, T., & Sage, D. (1996). Rhenium and molybdenum enrichments in sediments as indicators of oxic, suboxic and sulfidic conditions of deposition. *Earth and Planetary Science Letters*, 145(1–4), 65–78.

Dahl, T. W., Anbar, A. D., Gordon, G. W., et al. (2010). The behavior of molybdenum and its isotopes across the chemocline and in the sediments of sulfidic Lake Cadagno, Switzerland. *Geochimica et Cosmochimica Acta*, 74 (1), 144–163.

Dahl, T. W., Chappaz, A., Fitts, J. P., & Lyons, T. W. (2013). Molybdenum reduction in a sulfidic lake: Evidence from X-ray absorption fine-structure spectroscopy and implications for the Mo paleoproxy. *Geochimica et Cosmochimica Acta*, 103, 213–231.

Dahl, T. W., Chappaz, A., Hoek, J., et al. (2017). Evidence of molybdenum association with particulate organic matter under sulfidic conditions. *Geobiology*, 15(2), 311–323.

Dahl, T. W., & Wirth, S. B. (2017). Molybdenum isotope fractionation and speciation in a euxinic lake – Testing ways to discern isotope fractionation processes in a sulfidic setting. *Chemical Geology*, 460, 84–92.

Dickson, A. J. (2017). A molybdenum-isotope perspective on Phanerozoic deoxygenation events. *Nature Geoscience*, 10(10), 721–726.

Dickson, A. J., Idiz, E., Porcelli, D., & van den Boorn, S. H. (2019). The influence of thermal maturity on the stable isotope compositions and concentrations of molybdenum, zinc and cadmium in organic-rich marine mudrocks. *Geochimica et Cosmochimica Acta*, 1–16. DOI: https://doi.org/10.1016/j.gca.2019.11.001.

Emerson, S. R., & Huested, S. S. (1991). Ocean anoxia and the concentrations of molybdenum and vanadium in seawater. *Marine Chemistry*, 34(3–4), 177–196.

Erickson, B. E., & Helz, G. R. (2000). Molybdenum (VI) speciation in sulfidic waters: Stability and lability of thiomolybdates. *Geochimica et Cosmochimica Acta*, 64(7), 1149–1158.

Erickson, R. L. (1973). Crustal abundance of elements, and mineral reserves and resources. *US Geological Survey Professional Paper 820*, 21–25.

Glass, J. B., Wolfe-Simon, F., & Anbar, A. D. (2009). Coevolution of metal availability and nitrogen assimilation in cyanobacteria and algae. *Geobiology*, 7(2), 100–123.

Glass, J. B., Chappaz, A., Eustis, B., et al. (2013). Molybdenum geochemistry in a seasonally dysoxic Mo-limited lacustrine ecosystem. *Geochimica et Cosmochimica Acta*, 114, 204–219.

Goldberg, S., & Forster, H. S. (1998). Factors affecting molybdenum adsorption by soils and minerals. *Soil Science*, 163(2), 109–114.

Goldberg, S., Forster, H. S., & Godfrey, C. L. (1996). Molybdenum adsorption on oxides, clay minerals, and soils. *Soil Science Society of America Journal*, 60(2), 425–432.

Goldberg, T., Archer, C., Vance, D., & Poulton, S. W. (2009). Mo isotope fractionation during adsorption to Fe (oxyhydr)oxides. *Geochimica et Cosmochimica Acta*, 73(21), 6502–6516.

Goldschmidt, V. M. (1954). *Geochemistry*. Clarendon Press, Oxford.

Greaney, A. T., Rudnick, R. L., Romaniello, S. J., et al. (2020). Molybdenum isotope fractionation in glacial diamictites tracks the onset of oxidative weathering of the continental crust. *Earth and Planetary Science Letters*, 534, 116083.

Greber, N. D., Puchtel, I. S., Nägler, T. F., & Mezger, K. (2015). Komatiites constrain molybdenum isotope composition of the Earth's mantle. *Earth and Planetary Science Letters*, 421, 129–138.

Hardisty, D. S., Lyons, T. W., Riedinger, N., et al. (2018). An evaluation of sedimentary molybdenum and iron as proxies for pore fluid paleoredox conditions. *American Journal of Science*, 318(5), 527–556.

Helz, G. R., Bura-Nakić, E., Mikac, N., & Ciglenečki, I. (2011). New model for molybdenum behavior in euxinic waters. *Chemical Geology*, 284(3–4), 323–332.

Helz, G. R., Miller, C. V., Charnock, J. M., et al. (1996). Mechanism of molybdenum removal from the sea and its concentration in black shales: EXAFS evidence. *Geochimica et Cosmochimica Acta*, 60(19), 3631–3642.

Helz, G. R., & Vorlicek, T. P. (2019). Precipitation of molybdenum from euxinic waters and the role of organic matter. *Chemical Geology*, 509, 178–193.

Holland, H. D. (1984). *The chemical evolution of the atmosphere and oceans*. Princeton University Press.

Hutchings, A. M., Basu, A., Dickson, A. J., & Turchyn, A. V. (2020). Molybdenum geochemistry in salt marsh pond sediments. *Geochimica et Cosmochimica Acta*, 284, 75–91.

Kashiwabara, T., Takahashi, Y., Tanimizu, M., & Usui, A. (2011). Molecular-scale mechanisms of distribution and isotopic fractionation of molybdenum between seawater and ferromanganese oxides. *Geochimica et Cosmochimica Acta*, 75(19), 5762–5784.

Kendall, B., Dahl, T. W., & Anbar, A. D. (2017). The stable isotope geochemistry of molybdenum. *Reviews in Mineralogy and Geochemistry*, 82(1), 683–732.

Kendall, B., Komiya, T., Lyons, T. W., et al. (2015). Uranium and molybdenum isotope evidence for an episode of widespread ocean oxygenation during the late Ediacaran Period. *Geochimica et Cosmochimica Acta*, 156, 173–193.

Kerl, C. F., Lohmayer, R., Bura-Nakić, E., Vance, D., & Planer-Friedrich, B. (2017). Experimental confirmation of isotope fractionation in thiomolybdates using ion chromatographic separation and detection by multicollector ICPMS. *Analytical Chemistry*, 89(5), 3123–3129.

King, E. K., Perakis, S. S., & Pett-Ridge, J. C. (2018). Molybdenum isotope fractionation during adsorption to organic matter. *Geochimica et Cosmochimica Acta*, 222, 584–598.

King, E. K., & Pett-Ridge, J. C. (2018). Reassessing the dissolved molybdenum isotopic composition of ocean inputs: The effect of chemical weathering and groundwater. *Geology*, 46(11), 955–958.

Kuroda, P. K., & Sandell, E. B. (1954). Geochemistry of molybdenum. *Geochimica et Cosmochimica Acta*, 6(1), 35–63.

Liang, Liyuan, Romano Rinaldi, and Helmut Schober, eds. *Neutron applications in earth, energy and environmental sciences*. Springer Science & Business Media, 2008.

Liermann, L. J., Guynn, R. L., Anbar, A., & Brantley, S. L. (2005). Production of a molybdophore during metal-targeted dissolution of silicates by soil bacteria. *Chemical Geology*, 220(3–4), 285–302.

Lyons, T. W., Anbar, A. D., Severmann, S., Scott, C., & Gill, B. C. (2009). Tracking euxinia in the ancient ocean: A multiproxy perspective and Proterozoic case study. *Annual Review of Earth and Planetary Sciences*, 37, 507–534.

Lyons, T. W., Reinhard, C. T., & Planavsky, N. J. (2014). The rise of oxygen in Earth's early ocean and atmosphere. *Nature*, 506(7488), 307–315.

Miller, C. A., Peucker-Ehrenbrink, B., Walker, B. D., & Marcantonio, F. (2011), Re-assessing the surface cycling of molybdenum and rhenium. *Geochimica et Cosmochimica Acta*, 75, 7146–7179.

Nägler, T. F., Neubert, N., Böttcher, M. E., Dellwig, O., & Schnetger, B. (2011). Molybdenum isotope fractionation in pelagic euxinia: Evidence from the modern Black and Baltic Seas. *Chemical Geology*, 289(1–2), 1–11.

Nägler, T. F., Siebert, C., Lüschen, H., & Böttcher, M. E. (2005). Sedimentary Mo isotope record across the Holocene fresh–brackish water transition of the Black Sea. *Chemical Geology*, 219(1–4), 283–295.

Nakagawa, Y., Takano, S., Firdaus, M. L., et al. (2012). The molybdenum isotopic composition of the modern ocean. *Geochemical Journal*, 46(2), 131–141.

Neubert, N., Nägler, T. F., & Böttcher, M. E. (2008). Sulfidity controls molybdenum isotope fractionation into euxinic sediments: Evidence from the modern Black Sea. *Geology*, 36(10), 775–778.

Nissenbaum, A., & Swaine, D. J. (1976). Organic matter-metal interactions in recent sediments: The role of humic substances. *Geochimica et Cosmochimica Acta*, 40(7), 809–816.

Noordmann, J., Weyer, S., Montoya-Pino, C., Dellwig, O., Neubert, N., Eckert, S., Paetzel, M. and Böttcher, M., 2015, Uranium and molybdenum isotope systematics in modern euxinic basins: Case studies from the central Baltic Sea and the Kyllaren fjord (Norway): *Chemical Geology*, v. 396, p. 182–195, http://dx.doi.org/10.1016/j.chemgeo.2014.12.012

Pearce, C. R., Cohen, A. S., Coe, A. L., & Burton, K. W. (2008). Molybdenum isotope evidence for global ocean anoxia coupled with perturbations to the carbon cycle during the Early Jurassic. *Geology*, 36(3), 231–234.

Pedersen, T. F., Waters, R. D., & Macdonald, R. W. (1989). On the natural enrichment of cadmium and molybdenum in the sediments of Ucluelet Inlet, British Columbia. *Science of the Total Environment*, 79(2), 125–139.

Poulsón Brucker, R. L., McManus, J., Severmann, S., & Berelson, W. M. (2009). Molybdenum behavior during early diagenesis: Insights from Mo isotopes. *Geochemistry, Geophysics, Geosystems*, 10(6).

Proemse, B. C., Grasby, S. E., Wieser, M. E., Mayer, B., & Beauchamp, B. (2013). Molybdenum isotopic evidence for oxic marine conditions during the latest Permian extinction. *Geology*, 41(9), 967–970.

Rahaman, W., Singh, S. K., & Raghav, S. (2010). Dissolved Mo and U in rivers and estuaries of India: Implication to geochemistry of redox sensitive elements and their marine budgets. *Chemical Geology*, 278(3–4), 160–172.

Reich, M., Deditius, A., Chryssoulis, S., et al. (2013). Pyrite as a record of hydrothermal fluid evolution in a porphyry copper system: A SIMS/EMPA trace element study. *Geochimica et Cosmochimica Acta*, 104, 42–62.

Reimann, C., & De Caritat, P. (2012). *Chemical elements in the environment: Factsheets for the geochemist and environmental scientist*. Springer Science & Business Media.

Reinhard, C. T., Planavsky, N. J., Robbins, L. J., et al. (2013). Proterozoic ocean redox and biogeochemical stasis. *Proceedings of the National Academy of Sciences*, 110(14), 5357–5362.

Reitz, A., Wille, M., Nägler, T. F., & de Lange, G. J. (2007). Atypical Mo isotope signatures in eastern Mediterranean sediments. *Chemical Geology*, 245(1–2), 1–8.

Ross, S., & Sussman, A. (1955). Surface oxidation of molybdenum disulfide. *The Journal of Physical Chemistry*, 59(9), 889–892.

Scholz, F., Baum, M., Siebert, C., et al. (2018). Sedimentary molybdenum cycling in the aftermath of seawater inflow to the intermittently euxinic Gotland Deep, Central Baltic Sea. *Chemical Geology*, 491, 27–38.

Scholz, F., Siebert, C., Dale, A. W., & Frank, M. (2017). Intense molybdenum accumulation in sediments underneath a nitrogenous water column and implications for the reconstruction of paleo-redox conditions based on molybdenum isotopes. *Geochimica et Cosmochimica Acta*, 213, 400–417.

Scott, C., & Lyons, T. W. (2012). Contrasting molybdenum cycling and isotopic properties in euxinic versus non-euxinic sediments and sedimentary rocks: Refining the paleoproxies. *Chemical Geology*, 324, 19–27.

Scott, C., Lyons, T. W., Bekker, A., et al. (2008). Tracing the stepwise oxygenation of the Proterozoic ocean. *Nature*, 452(7186), 456–459.

Seralathan, P., & Hartmann, M. (1986). Molybdenum and vanadium in sediment cores from the NW-African continental margin and their relations to climatic and environmental conditions. *Meteor-Forschungsergebnisse. Reihe C, Geologie und Geophysik*, (40), 1–17.

Shaw, T. J., Gieskes, J. M., & Jahnke, R. A. (1990). Early diagenesis in differing depositional environments: The response of transition metals in pore water. *Geochimica et Cosmochimica Acta*, 54(5), 1233–1246.

Siebert, C., Nägler, T. F., von Blanckenburg, F., & Kramers, J. D. (2003). Molybdenum isotope records as a potential new proxy for paleoceanography. *Earth and Planetary Science Letters*, 211(1–2), 159–171.

Siebert, C., Kramers, J. D., Meisel, T., Morel, P., & Nägler, T. F. (2005). PGE, Re-Os, and Mo isotope systematics in Archean and early Proterozoic sedimentary systems as proxies for redox conditions of the early Earth. *Geochimica et Cosmochimica Acta*, 69(7), 1787–1801.

Smedley, P. L., & Kinniburgh, D. G. (2017). Molybdenum in natural waters: A review of occurrence, distributions, and controls. *Applied Geochemistry*, 84, 387–432.

Sweere, T. C., Hennekam, R., Vance, D., & Reichart, G. J. (2021). Molybdenum isotope constraints on the temporal development of sulfidic conditions during Mediterranean sapropel intervals. *Geochemical Perspectives Letters*, 17, 16–20.

Szilagyi, M. (1967). *Sorption of molybdenum by humus preparations*. Nuclear Research Institute, Debrecen, Hungary.

Tessin, A., Chappaz, A., Hendy, I., & Sheldon, N. (2019). Molybdenum speciation as a paleo-redox proxy: A case study from Late Cretaceous Western Interior Seaway black shales. *Geology*, 47(1), 59–62.

Tossell, J. A. (2005). Calculating the partitioning of the isotopes of Mo between oxidic and sulfidic species in aqueous solution. *Geochimica et Cosmochimica Acta*, 69, 2981–2993.

Tribovillard, N., Algeo, T. J., Lyons, T., & Riboulleau, A. (2006). Trace metals as paleoredox and paleoproductivity proxies: An update. *Chemical Geology*, 232(1–2), 12–32.

Turekian, K. K., & Wedepohl, K. H. (1961). Distribution of the elements in some major units of the earth's crust. *Geological Society of America Bulletin*, 72(2), 175–192.

Voegelin, A. R., Pettke, T., Greber, N. D., von Niederhäusern, B., & Naegler, T. F. (2014). Magma differentiation fractionates Mo isotope ratios: Evidence from the Kos Plateau Tuff (Aegean Arc). *Lithos*, 190, 440–448.

Vorlicek, T. P., Helz, G. R., Chappaz, A., et al. (2018). Molybdenum burial mechanism in sulfidic sediments: iron-sulfide pathway. *ACS Earth and Space Chemistry*, 2(6), 565–576.

Wagner, M. A., Chappaz, and T. W. Lyons. (2017). Molybdenum speciation and burial pathway in weakly sulfidic environments: Insights from XAFS. *Geochimica et Cosmochimica Acta*, 206, 18–29.

Wang, D., Aller, R. C., & Sañudo-Wilhelmy, S. A. (2011). Redox speciation and early diagenetic behavior of dissolved molybdenum in sulfidic muds. *Marine Chemistry*, 125(1–4), 101–107.

Wasylenki, L. E., Anbar, A. D., Liermann, L. J., et al. (2007). Isotope fractionation during microbial metal uptake measured by MC-ICP-MS. *Journal of Analytical Atomic Spectrometry*, 22(8), 905–910.

Wasylenki, L. E., Rolfe, B. A., Weeks, C. L., Spiro, T. G., & Anbar, A. D. (2008). Experimental investigation of the effects of temperature and ionic strength on Mo isotope fractionation during adsorption to manganese oxides. *Geochimica et Cosmochimica Acta*, 72(24), 5997–6005.

Wasylenki, L. E., Weeks, C. L., Bargar, J. R., et al. (2011). The molecular mechanism of Mo isotope fractionation during adsorption to birnessite. *Geochimica et Cosmochimica Acta*, 75(17),5019–5031.

Wedepohl, K. H. (1968). Chemical fractionation in the sedimentary environment. In L. H. Ahrens (ed.), *Origin and Distribution of the Elements* (pp. 999–1016). Pergamon.

Wedepohl, K. H. (1971). Environmental influences on the chemical composition of shales and clays. *Physics and Chemistry of the Earth*, 8, 307–333.

Wichard, T., Mishra, B., Myneni, S. C., Bellenger, J. P., & Kraepiel, A. M. (2009). Storage and bioavailability of molybdenum in soils increased by organic matter complexation. *Nature Geoscience*, 2(9), 625–629.

Wille, M., Kramers, J. D., Nägler, T. F., et al. (2007). Evidence for a gradual rise of oxygen between 2.6 and 2.5 Ga from Mo isotopes and Re-PGE signatures in shales. *Geochimica et Cosmochimica Acta*, 71(10), 2417–2435.

Wille, M., Nägler, T. F., Lehmann, B., Schröder, S., & Kramers, J. D. (2008). Hydrogen sulphide release to surface waters at the Precambrian/Cambrian boundary. *Nature*, 453 (7196), 767–769.

Zerkle, A. L., Scheiderich, K., Maresca, J. A., Liermann, L. J., & Brantley, S. L. (2011). Molybdenum isotope fractionation by cyanobacterial assimilation during nitrate utilization and N2 fixation. *Geobiology*, 9(1), 94–106.

Zheng, Y., Anderson, R. F., Van Geen, A., & Kuwabara, J. (2000). Authigenic molybdenum formation in marine sediments: A link to pore water sulfide in the Santa Barbara Basin. *Geochimica et Cosmochimica Acta*, 64(24), 4165–4178.

Zhou, L., Algeo, T. J., Shen, J., et al. (2015). Changes in marine productivity and redox conditions during the Late Ordovician Hirnantian glaciation. *Palaeogeography, Palaeoclimatology, Palaeoecology*, 420, 223–234.

# Cambridge Elements ≡

# Geochemical Tracers in Earth System Science

## Timothy Lyons
### University of California
Timothy Lyons is a Distinguished Professor of Biogeochemistry in the Department of Earth Sciences at the University of California, Riverside. He is an expert in the use of geochemical tracers for applications in astrobiology, geobiology and Earth history. Professor Lyons leads the 'Alternative Earths' team of the NASA Astrobiology Institute and the Alternative Earths Astrobiology Center at UC Riverside.

## Alexandra Turchyn
### University of Cambridge
Alexandra Turchyn is a University Reader in Biogeochemistry in the Department of Earth Sciences at the University of Cambridge. Her primary research interests are in isotope geochemistry and the application of geochemistry to interrogate modern and past environments.

## Chris Reinhard
### Georgia Institute of Technology
Chris Reinhard is an Assistant Professor in the Department of Earth and Atmospheric Sciences at the Georgia Institute of Technology. His research focuses on biogeochemistry and paleoclimatology, and he is an Institutional PI on the 'Alternative Earths' team of the NASA Astrobiology Institute.

## About the Series

This innovative series provides authoritative, concise overviews of the many novel isotope and elemental systems that can be used as 'proxies' or 'geochemical tracers' to reconstruct past environments over thousands to millions to billions of years – from the evolving chemistry of the atmosphere and oceans to their cause-and-effect relationships with life. Covering a wide variety of geochemical tracers, the series reviews each method in terms of the geochemical underpinnings, the promises and pitfalls, and the 'state-of-the-art' and future prospects, providing a dynamic reference resource for graduate students, researchers and scientists in geochemistry, astrobiology, paleontology, paleoceanography and paleoclimatology.

The short, timely, broadly accessible papers provide much-needed primers for a wide audience – highlighting the cutting-edge of both new and established proxies as applied to diverse questions about Earth system evolution over wide-ranging time scales.

Cambridge Elements ⁼

# Geochemical Tracers in Earth System Science

### Elements in the Series

A full series listing is available at: www.cambridge.org/EESS